Benchmark Tests on the
Generation of Fair Shapes
subject to Constraints

I. Applegarth / P. D. Kaklis / S. Wahl (Eds.)

Benchmark Tests on the Generation of Fair Shapes subject to Constraints

Edited by
I. Applegarth, Kockums Computer Systems UK, Ltd.
P. D. Kaklis, National Technical University of Athens
S. Wahl, Daimler-Benz AG

based on
the Benchmarking Experiment
organized by the HCM-Network FAIRSHAPE
ERB-CHRXCT-940522
Spring-Summer 1997

 B. G. Teubner Stuttgart · Leipzig 2000

Die Deutsche Bibliothek – CIP-Einheitsaufnahme

Ein Titeldatensatz für diese Publikation ist bei
Der Deutschen Bibliothek erhältlich

Das Werk einschließlich aller seiner Teile ist urheberrechlich geschützt. Jede Verwertung außerhalb der engen Grenzen des Urheberrechtsgesetzes ist ohne Zustimmung des Verlages unzulässig und strafbar. Das gilt besonders für Vervielfältigungen, Übersetzungen, Mikroverfilmungen und die Einspeicherung und Verarbeitung in elektronischen Systemen.

© 2000 B. G. Teubner Stuttgart · Leipzig
Printed in Germany
Druck und Binden: Druck Partner Rübelmann GmbH, Hemsbach

Preface

According to the approved Work Programme of the **H**uman **C**apital & **M**obility (**HCM**) Project FAIRSHAPE (ERB-CHRXCT-940522), the project was planning to conclude its activities *"with a Benchmarking Workshop, presenting and testing the methods developed in the context of the network"*. The above commitment was re-confirmed during the 4th Internal Workshop of FAIRSHAPE (Athens, 21-22.02.97), and a sub-committee was appointed for conducting a Benchmarking Experiment.

The scientific scope of the Benchmark Experiment was to establish a state-of-the-art evaluation of the quality of fairness, achievable by current curve and surface fairing and shape-preserving methodologies in the context of alternative fairness measures and geometric constraints. Since it has been decided that the Benchmark Experiment should not be limited within FAIRSHAPE, the scope of the initiative and the participation rules were distributed by e-mail, in early spring '97, to an extensive list of European researchers that are active in the area of **C**omputer-**A**ided **G**eometric **D**esign (**CAGD**). As a result of this, rather limited, publicity, two (2) of the eventually received contributions originated from research groups not belonging to FAIRSHAPE. Overall, the current Benchmark Collection comprises eight (8) contributions, two (2) of which originated from industry:

1. **Daimler-Benz AG** - FAIRSHAPE partner, senior scientist: E. Kaufmann, young scientist: S. Wahl,

2. **Kockums Computer Systems UK, Ltd.** - FAIRSHAPE partner, senior scientist: I. Applegarth,

while the remaining six (6) originated from Academia:

1. **University of Kaiserslautern**, senior scientist: H. Hagen, young scientists: M. Latz and S. Heinz,

2. **University of Erlangen-Nürnberg**, senior scientist: G. Greiner, young scientist: K. Hormann,

3. **Technical University of Berlin** - FAIRSHAPE partner, senior scientist: H. Nowacki, young scientists: J. Heimann and G. Westgaard,

4. **Technical University of Darmstadt** - FAIRSHAPE partner, senior scientist: J. Hoschek, young scientists: U. Dietz and J. Hadenfeld,

5. **University of Florence** - FAIRSHAPE partner, senior scientists: F. Fontanella, P. Costantini, C. Manni, young scientist: S. Asaturyan,

6. **National Technical University of Athens** - FAIRSHAPE partner, senior scientist: P.D. Kaklis, young scientists: N.C. Gabrielides, G.D. Koras, K.G. Pigounakis and T.P. Gerostathis.

The deadline for contributing to the Benchmark Experiment was July 15th, 1997. A first presentation of the Benchmark Collection was offered to the European CAGD community at the opening day of the International Benchmarking Workshop on *"Creating Fair and Shape-Preserving Curves and Surfaces"*, organized by FAIRSHAPE (local organizer: Prof.

Dr.-Ing. H. Nowacki, Technical University of Berlin) and held in Kleinmachnow (near Berlin and Potsdam) on September 14-17, 1997. The workshop has been attended by more than fifty (50) participants and a set of twenty-one (21) significant papers was selected, after thorough review, for publication in a volume entitled: *"Creating Fair and Shape-Preserving Curves and Surfaces"*, H. Nowacki and P.D. Kaklis (Eds.), B.G. Teubner Stuttgart, 1998, 288 pages.

The principal scope of the present volume is to provide the international CAGD community with a representative sample from the material collected through the FAIRSHAPE Benchmarking Experiment. This sample is contained in three (3) consecutive sections, focusing on:

- **Shape-Preserving Interpolation** (§2), presenting two (2) contributions from benchmarking with the three-dimensional *chair* point-set,

- **Fairing** (§3), presenting three (3) contributions from benchmarking with the *chine* curve from a fast boat and five (5) contributions from benchmarking with a patch from the *hood* of a car-model, and

- **Shape-Constrained Approximation** (§4), presenting three (3) contributions from benchmarking with measurements from the surface of a *lens* and two (2) contributions from benchmarking with a set of *waterLines* near the fore part of a ship.

Section 5 attempts to comparatively evaluate the quality of the results collected in §§2-4.

Finally, §6 provides the FINAL ANNOUNCEMENT of the Benchmark Experiment, containing the GENERAL RULES OF THE BENCHMARK and all necessary information for downloading the benchmark data and the fairness-evaluation software as well as uploading future contributions to the proper ftp-site.

On this occasion, we would like to stress that the FAIRSHAPE Benchmarking Experiment is still open to the interested reader-researcher. The full Benchmark Collection, as well as future contributions, will be made available to the scientific community via a Web page that is currently under construction and will be located at the home page of the Department of Mathematics of the Technical University of Darmstadt (URL: http://www.mathematik.th-darmstadt.de). This task has been undertaken mainly by Prof. J. Hoschek and Dr. J. Hadenfeld, TU Darmstadt, whose initiative, being materialized after the official end of FAIRSHAPE, is gratefully appreciated.

In fine, we would like to cordially thank Prof. Dr.-Ing. H. Nowacki, TU Berlin, for his generous support and guidance offered during all phases of the FAIRSHAPE Benchmarking Experiment, from setting its rules until composing the present volume. Sincere thanks are also due to Dr.-Ing K.G. Pigounakis and Dipl.-Ing A.I. Ginnis, TU Athens, for preparing and testing the fairness-evaluation software, as well as Dipl.-Ing. N.C. Gabrielides, Dipl.-Ing. G. Koras and Dipl.-Ing. K. Kostas, TU Athens, for typing in LaTeX and organizing the graphic contents of this volume.

Last but not least, the financial and administrative support extended to FAIRSHAPE by the European Commission and, especially, the collaborative spirit of the responsible DG-XII-G.2 officer, Mr. Ralph Rahders, are gratefully acknowledged.

Ian Applegarth
Panagiotis D. Kaklis
Steffen Wahl
(Co-editors)

CONTENTS

                                                                                                                              page

1. Overview 9
2. Benchmark in the Area of Shape Preserving Interpolation 12
2.1 Chair - Benchmark 12
2.1.1 Problem (INIT) 12
2.1.2 National Technical University of Athens (NTUA) 13
2.1.3 University of Florence (U_FIR) 15
2.1.4 Summary 16
3. Benchmark in the Area of Fairing 17
3.1 Chine3 - Benchmark 17
3.1.1 Problem (INIT) 17
3.1.2 National Technical University of Athens (NTUA) 18
3.1.3 Technical University of Darmstadt; Hadenfeld (TUDH) 19
3.1.4 Kockums Computer Systems (KCS)UK, Ltd 23
3.1.5 Summary 25
3.2 Hood - Benchmark 26
3.2.1 Problem (INIT) 26
3.2.2 National Technical University of Athens (NTUA) 26
3.2.3 Technical University of Berlin (TUB) 27
3.2.4 Technical University of Darmstadt; Dietz (TUDD) 29
3.2.5 Technical University of Darmstadt; Hadenfeld (TUDH) 30
3.2.6 Daimler-Benz AG (DBAG) 30
3.2.7 Kockums Computer Systems (KCS)UK, Ltd 31
3.2.8 Summary 32
4. Benchmark in the Area of Shape - Constrained Approximation 33
4.1 Lens - Benchmark 33
4.1.1 Problem (INIT) 33
4.1.2 Technical University of Darmstadt (TUD) 33
4.1.3 University of Erlangen-Nürnbreg (U_EN) 34
4.1.4 University of Kaiserslautern (U_K) 35
4.1.5 Summary 36
4.2 WaterLines - Benchmark 37
4.2.1 Problem (INIT) 37
4.2.2 University of Erlangen-Nürnbreg (U_EN) 37
4.2.3 University of Kaiserslautern (U_K) 39
4.2.4 Summary 40
5. Comparative Remarks 41
6. Doing It Yourself 43
Appendix: Colour Plates 49
References 65

1 Overview

The material contained in the present volume is structured in six (6) sections and an Appendix, containing colour plates. In the sequel of the current overview, sections 2-4 constitute the kernel of the book, providing the reader with a systematic presentation of a representative subset from the received contributions to the FAIRSHAPE Benchmarking Experiment. These sections conform with the classification of the Benchmark Problems in three (3) major groups, namely:

1. **Benchmark in the Area of Shape-Preserving Interpolation (§2)**, presenting contributions from benchmarking with the three-dimensional *chair* point-set; see Fig. 2.1.1.

For the **Chair-Benchmark**, contributions have been received from:

- the National Technical University of Athens (§2.1.2) and
- the University of Florence (§2.1.3).

2. **Benchmark in the Area of Fairing (§3)**, presenting contributions from benchmarking with the *chine curve* of a fast boat (see Fig 3.1.1) and a surface patch measured from the *hood* of a car-model; see Fig. 3.2.1 in the Appendix: Colour Plates.

For the **Chine3-Benchmark**, contributions have been received from:

- the National Technical University of Athens (§3.1.2),
- the Technical University of Darmstadt (§3.1.3), and
- Kockums Computer Systems UK Ltd (§3.1.4).

For the **Hood-Benchmark**, contributions have been received from:

- the National Technical University of Athens (§3.2.2),
- the Technical University of Berlin (§3.2.3),
- the Technical University of Darmstadt (§§3.2.4-5),
- Daimler - Benz AG (§3.2.6), and
- Kockums Computer Systems UK, Ltd (§3.2.7).

3. **Benchmark in the Area of Shape-Constrained Approximation (§4)**, presenting the contributions from benchmarking with measurements from the surface of a *lens* (see Fig. 4.1.1) and a set of *waterlines* near the fore part of a ship; see Fig. 4.2.1.

For the **Lens - Benchmark**, contributions have been received from:

- the Technical University of Darmstadt (§4.1.2),
- the University of Erlangen-Nürnberg (§4.1.3) and
- the University of Kaiserslautern (§4.1.4).

For the **WaterLines-Benchmark**, contributions have been received from:

- the University of Erlangen-Nürnberg (§4.2.2) and

- the University of Kaiserslautern (§§4.2.3-4).

Each benchmark problem is documented with a description of the input data, in text and graphics format, and the associated constraints; see, e.g., §3.1.1. Next, the documentation includes the contribution of each participant, structured as follows:

- the name and address of the scientists that have been involved in the experiment,
- a very brief description of the employed technique, usually pointing to the pertinent reference(s) of the literature, and
- the values of a list of functionals, that measure the fairness of a curve or surface. These functionals are listed below:

(i) Let $\mathbf{Q}(u)$, $u \in I = [u_1, u_N]$, be a sufficiently smooth parametric curve in the two/three-dimensional affine space $I\!\!E^2/I\!\!E^3$, with s, κ and τ denoting its arc length, curvature and torsion distributions, respectively. Furthermore, let $\|\bullet\|$ denote the Euclidean norm in the working affine space. The functionals, that have been adopted by the Steering Committee of FAIRSHAPE for measuring the fairness of a curve, have as follows:

$L_{21} = \int_I \|\frac{d\mathbf{Q}(u)}{du}\|^2 du,$

$L_{22} = \int_I \|\frac{d^2\mathbf{Q}(u)}{du^2}\|^2 du,$

$L_{23} = \int_I \|\frac{d^3\mathbf{Q}(u)}{du^3}\|^2 du,$

$L_{2\kappa} = \int_I \kappa^2(s) ds,$

$L_{2\kappa'} = \int_I \left(\frac{d\kappa(s)}{ds}\right)^2 ds,$

$L_{2\tau} = \int_I \tau^2(s) ds,$

max_κ: maximum of the absolute of curvature over I,

$\#mono\_\kappa$: number of monotonic segments in curvature distribution,

max_τ: maximum of the absolute torsion over I,

$\#sign\_\tau$: number of sign changes in torsion distribution.

(ii) Let $\mathbf{S}(u,v)$, $(u,v) \in \Omega = [u_1, u_N] \times [v_1, v_M]$, be a sufficiently smooth parametric surface, with κ_i, $i = 1, 2$, denoting its principal curvatures and the subscript u/v indicating partial differentiation with respect to u/v, respectively. The functionals, that have been adopted by the Steering Committee of FAIRSHAPE for measuring the fairness of a surface, have as follows:

$L_{22} = \int_\Omega (\|\mathbf{S}_{uu}\|^2 + 2\|\mathbf{S}_{uv}\|^2 + \|\mathbf{S}_{vv}\|^2) du dv,$

$L_{23} = \int_\Omega (\|\mathbf{S}_{uuv}\|^2 + \|\mathbf{S}_{vvu}\|^2) du dv,$

$L_{2\kappa_{1,2}} = \int_\Omega (\kappa_1^2 + \kappa_2^2) d\Omega,$

$max|\kappa_{1+2}| : \max_\Omega(|\kappa_1| + |\kappa_2|),$

$maxK$: maximum of the absolute of Gaussian curvature K over Ω,

$maxH$: maximum of the absolute of mean curvature over Ω.

The documentation concludes with graphical information, usually containing curvature and torsion plots, for curves, and Gaussian-, mean-curvature as well as isophote distributions for surfaces. More specifically, the Gaussian- and mean-curvature distributions are given in the form of colour plates, all collected within the homonymous Appendix.

Note that, by anonymous ftp at `ftp.deslab.naval.ntua.gr`, which is the address of the ftp-site of the Ship-Design Laboratory of the National Technical University of Athens, the perspective benchmarker can download a set of FORTRAN subroutines, located in the directories:
`pub/deslab/CAGD/FAIRSHAPE_curve_crit`
`pub/deslab/CAGD/FAIRSHAPE_surface_crit`,
and use them for evaluating the afore mentioned curve and surface fairness measures; see also §6. This can be done if the curve/surface under consideration is available in B-spline format or can be evaluated over a user-specified orthogonal grid of its parametric domain of definition.

Section 5 provides comparative remarks on the quality of the benchmark contributions collected in §§2-4.

Finally, §6 enables the interested reader-researcher to test her/his *fairing* and/or *shape-preservation* software against the data proposed by the FAIRSHAPE Benchmarking Experiment. For this purpose, §6 includes the FINAL ANNOUNCEMENT of the Benchmarking Experiment, containing the GENERAL RULES OF THE BENCHMARK, as well as all necessary information for downloading the required data/software or uploading a benchmark contribution to the proper `ftp-site`; see Appendices I-IV of the FINAL ANNOUNCEMENT.

2 Benchmark in the Area of Shape Preserving Interpolation

2.1 Chair - Benchmark

2.1.1 Problem (INIT)

2.1.1.1 Description The data set, supplied by NTUA, is a periodic point-set $\mathcal{D}$ consisting of thirteen (13) points $\mathbf{I}_m$, $m = 1(1)13$; see the rhombuses in Fig. 2.1.1. $\mathcal{D}$ is yz-symmetric and contains two distinct quadruples of coplanar points, namely the point-sets: $\{\mathbf{I}_m,\ m = 2(1)5\}$ and $\{\mathbf{I}_m,\ m = 9(1)12\}$, and one triplet of collinear points, namely the point-set: $\{\mathbf{I}_m,\ m = 6(1)8\}$.

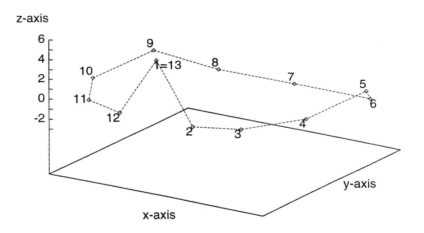

Figure 2.1.1 Chair - data (INIT)

2.1.1.2 Constraints Construct a curve $\mathbf{Q}(u)$ so that:

(i) $\mathbf{Q}(u)$ exhibits Frénet-frame continuity of order 3, i.e., continuity of unit-tangent and binormal vectors, curvature and torsion distributions,

(ii) $\mathbf{Q}(u)$ interpolates $\mathcal{D}$ with a user-specified parameterization $\mathcal{U}$,

(iii) $\mathbf{Q}(u)$ is shape-preserving in the following sense: Let $\mathbf{L}_m = \mathbf{I}_{m+1} - \mathbf{I}_m$, $m = 0(1)N$ ($\mathbf{I}_0 = \mathbf{I}_{N-1}$, $\mathbf{I}_{N+1} = \mathbf{I}_2$), $\mathbf{P}_m = \mathbf{L}_{m-1} \times \mathbf{L}_m$, which will be referred to as the convexity indicators at the point $\mathbf{I}_m$, $\Delta_m = det\big([\mathbf{L}_{m-1}, \mathbf{L}_m, \mathbf{L}_{m+1}]\big)$ and $\mathbf{w}(u) = \dot{\mathbf{Q}}(u) \times \ddot{\mathbf{Q}}(u)$, with the dot denoting differentiation with respect to the parameter u. Then:

(iii.1) (Convexity criterion) If $\mathbf{P}_m \cdot \mathbf{P}_{m+1} > 0$, then:
$$\mathbf{w}(u) \cdot \mathbf{P}_n > 0, \quad u \in [u_m, u_{m+1}], \quad n = m, m+1.$$

(iii.2) (Torsion criterion) If $\Delta_m \neq 0$, then:
$$\tau(u)\Delta_m > 0, \quad u \in (u_m, u_{m+1}), \text{ and}$$
if $\Delta_m \Delta_{m+1} > 0$, then:
$$\tau(u)\Delta_n > 0, \quad n = m-1, m.$$

Furthermore, $\mathbf{Q}(u)$ should be as coplanar/collinear as possible in the corresponding parametric intervals, i.e., in $[u_2, u_5]$, $[u_9, u_{12}]$ for the coplanar and $[u_6, u_8]$ for the collinear case, while its maximum absolute curvature and torsion should be kept as small as possible.

2.1.2 National Technical University of Athens (NTUA)

2.1.2.1 Name and Address M.I. Karavelas, P.D. Kaklis,
Department of Naval Architecture and Marine Engineering,
National Technical University of Athens,
Athens, GREECE,
{menelaos,kaklis}@deslab.ntua.gr,
Date: 5.6.97

2.1.2.2 Method description The adopted method relies on the use of polynomial splines on non-uniform degree that can preserve the shape properties of the associated linear interpolant. The algorithm is described, in detail, in [11].

2.1.2.3 Numerical output of fairness measures

```
L21         .................................................  42.093640016
L22         .................................................  11.083656495
L23         .................................................  118.46251543
L2kappa     .................................................  433.87277481
L2kappa'    .................................................  30642550.192
L2tau       .................................................  61.501551225
max_kappa   .................................................  322.32606111
#mono_kappa .................................................  24
max_tau     .................................................  14.848353330
#sign_tau   .................................................  9
```

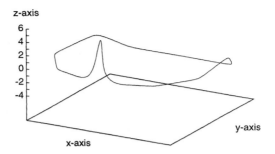

Figure 2.1.2a Chair - curve (NTUA)

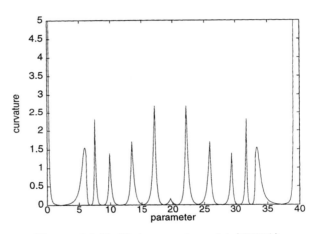

Figure 2.1.2b Chair - curvature plot (NTUA)

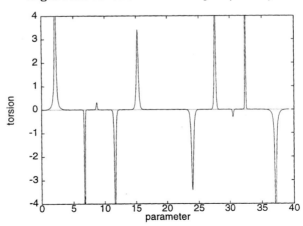

Figure 2.1.2c Chair - torsion plot (NTUA)

2.1.2.4 Graphical Output See Fig. 2.1.2.

2.1.3 University of Florence (U_FIR)

2.1.3.1 Name and Address Souren Asaturyan, Carla Manni
Department of Energetics
University of Florence
Florence, ITALY
{souren,manni}@ing.unifi.it

Paolo Costantini
Department of Mathematics
University of Siena
Siena, ITALY
costantini@unisi.it

2.1.3.2 Method description
The employed method is a local method, that provides shape preserving space curves by using sextic polynomial splines. The interpolating curve satisfies the convexity and torsion criteria associated to the polygonal interpolant. The continuity of the curvature vector and the torsion are obtained by appropriate projections and fulfillment of shape-preservation constraints.

2.1.3.3 Numerical output of fairness measures

L21	195.85698650
L22	1664.1110039
L23	66852.360902
L2kappa	77.949622962
L2kappa'	2.6721449134E+016
L2tau	6.0663970043
max_kappa	21.833191423
#mono_kappa	31
max_tau	3.0734983293
#sign_tau	8

2.1.3.4 Graphical Output
See Fig. 2.13. Note that Figs. 2.1.3c and 2.1.3d depict the distribution of the inner product, denoted as $< \cdot, \cdot >$, between the curvature vector and the convexity indicators $\mathbf{P}_i$ and $\mathbf{P}_{i+1}$, respectively.

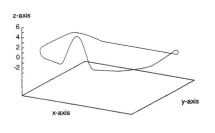

Figure 2.1.3a Chair - curve (U_FIR) **Figure 2.1.3b** Chair - torsion plot (U_FIR)

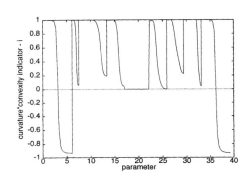
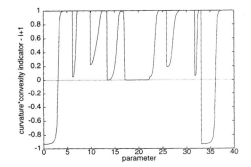

Figure 2.1.3c Chair - $< \mathbf{w}, \mathbf{P}_i >$ (U_FIR) **Figure 2.1.3d** Chair - $< \mathbf{w}, \mathbf{P}_{i+1} >$ (U_FIR)

2.1.4 SUMMARY

Participants

NTUA M.I. Karavelas, P.D. Kaklis
U_FIR S. Asaturyan, C. Manni, P. Costantini

Numerical output

	L_{21}	L_{22}	L_{23}	$L_{2\kappa}$	$L_{2\kappa'}$	$L_{2\tau}$
NTUA	42.094	11.08	118.5	433.87	3×10^7	61.5016
U_FIR	195.857	1664.11	66852.3	77.95	2×10^{16}	6.0664

	max_κ	$\#mono\_\kappa$	max_τ	$\#sign\_\tau$
NTUA	322.33	24	14.8483	9
U_FIR	21.83	31	3.0735	8

3 Benchmark in the Area of Fairing

3.1 Chine3 - Benchmark

3.1.1 Problem (INIT)

3.1.1.1 Description The curve represents the chine from the hull of a fast boat, and it is represented as a cubic B-Spline with fifteen (15) control points. Units are in meters. The data has been supplied by NTUA; see Fig. 3.1.1.

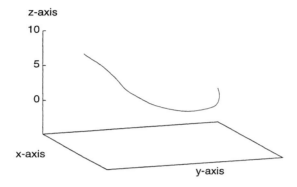

Figure 3.1.1a Chine3 - curve (INIT)

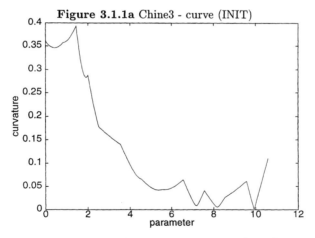

Figure 3.1.1b Chine3 - curvature plot (INIT)

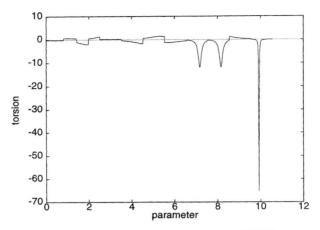

Figure 3.1.1c Chine3 - torsion plot (INIT)

3.1.1.2 Constraints At the ends of the curve, C^0 conditions should be maintained. The maximum deviation from the original curve should be less than 20mm.

3.1.1.3 Numerical output of fairness measures

```
L21      ..................................................  10.59440487005
L22      ..................................................  0.320156616125
L23      ..................................................  0.377561372135
L2kappa  ..................................................  0.317821223157
L2kappa' ..................................................  0.109989237980
L2tau    ..................................................  258.3511654492
max_kappa ..................................................  0.393390875878
#mono_kappa ..................................................  14
max_tau  ..................................................  134.6232779099
#sign_tau ..................................................  8
```

3.1.2 National Technical University of Athens (NTUA)

3.1.2.1 Name and Address K.G. Pigounakis, P.D. Kaklis
Department of Naval Architecture and Marine Engineering
National Technical University of Athens
Athens, GREECE
{kpig,kaklis}@deslab.ntua.gr
Date: 22.8.97

3.1.2.2 Method description The method is based on a knot-removal, knot-reinsertion technique, that is described, in detail, in Pigounakis et al. [12].

3.1.2.3 Numerical output of constraints

3.1.2.4 Numerical output of fairness measures

```
L21        ..........................................  10.58510164394
L22        ..........................................   0.2907278831813
L23        ..........................................   0.0535605413909
L2kappa    ..........................................   0.2938760099559
L2kappa'   ..........................................   0.0280257123041
L2tau      ..........................................  11.79808487573
max_kappa  ..........................................   0.4112133225768
#mono_kappa ..........................................   2
max_tau    ..........................................   7.478243721619
#sign_tau  ..........................................   4
```

3.1.2.5 Graphical Output   See Fig. 3.1.2.

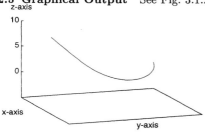
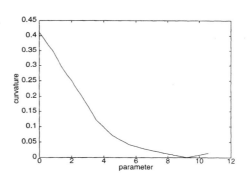

Figure 3.1.2a Chine3 - curve (NTUA)   **Figure 3.1.2b** Chine3 - curvature plot (NTUA)

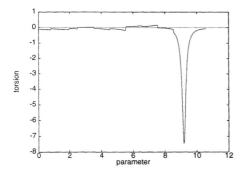
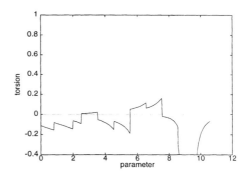

Figure 3.1.2c Chine3 - torsion plot (NTUA)   **Figure 3.1.2d** Chine3 - torsion plot (NTUA)

3.1.2.6 Comments
The employed hardware platform was a DEC 5000/200 with R2000 RISC processor and ULTRIX operating system. CPU-time: 3.10 sec.

3.1.3 Technical University of Darmstadt; Hadenfeld (TUDH)

3.1.3.1 Name and Address   Jan Hadenfeld, Josef Hoschek

Fachbereich Mathematik, AG3
Technical University of Darmstadt
Darmstadt, GERMANY
hadenfeld@mathematik.tu-darmstadt.de
Date: 9.7.1997

3.1.3.2 Method description
The faired curve has been obtained by minimizing the integral of the squared norm of the third parametric derivative of the curve.

3.1.3.3 Numerical output of constraints

3.1.3.4 Numerical output of fairness measures

```
L21        ............................................ 10.5919656
L22        ............................................  0.27896045
L23        ............................................  0.02865433
L2kappa    ............................................  0.27397734
L2kappa'   ............................................  0.02742591
L2tau      ............................................  0.65430279
max_kappa  ............................................  0.34459419
#mono_kappa ...................................  3
max_tau    ............................................  0.91691438
#sign_tau  ......................................  2
```

3.1.3.5 Graphical Output
See Fig. 3.1.3. In Figs 3.1.3a and 3.1.3b, the radius of the depicted circles is equal to the curvature, while porcupines indicate the direction and the length of the normals.

3 Benchmark in the Area of Fairing 21

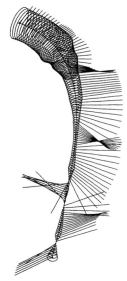

Figure 3.1.3a Chine3 - curve along with curvature tube and principal-normal porcupine plot (INIT)

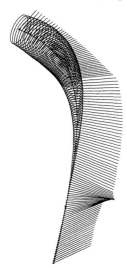

Figure 3.1.3b Chine3 - curve along with curvature tube and principal-normal porcupine plot (TUDH)

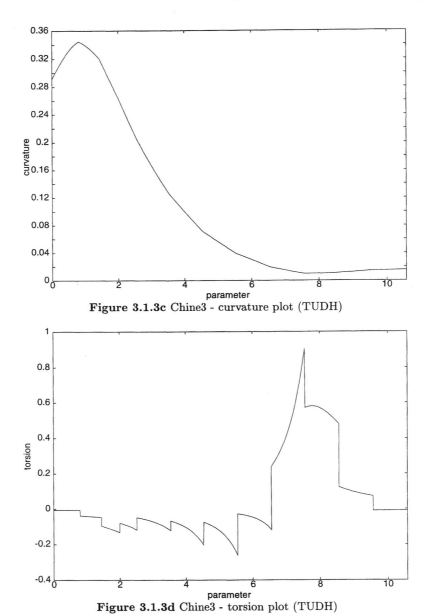

Figure 3.1.3c Chine3 - curvature plot (TUDH)

Figure 3.1.3d Chine3 - torsion plot (TUDH)

3.1.3.6 Comments The employed hardware platform was a PC with a Pentium-Pro processor at 200 MHz and LINUX operating system. Figures 3.1.3a and 3.1.3b are made with the aid of the commercial software package: *Surfacer* (Courtesy: IMAGEWARE).

3.1.4 Kockums Computer Systems (KCS) UK, Ltd

3.1.4.1 Name and Address Ian Applegarth
Kockums Computer Systems (KCS) UK, Ltd
UNITED KINGDOM
iana@kcs.co.uk

3.1.4.2 Method description Two methods have been employed:
Method 1: This is based on the method developed in Eck and Hadenfeld [2] and it will be herein denoted as the EH-method. This approach uses the minimization of the integral L_2 of the squared norm of the second parametric derivative of the curve.
Method 2: This is based on the method developed in Pigounakis et al [12] and it will be herein denoted as the PSK-method. The criteria for fairing are based on the following requirements:

- the curvature plot should be comprised of as few as possible monotonic segments,
- the torsion plot should be as close as possible to being continuous also with the fewest number of monotonic pieces,
- sign changes in the torsion plot should be as few as possible,
- the value of torsion, at each point of the curve, should be as small as possible.

3.1.4.3 Numerical output of constraints

- EH-method

```
D1 deviation from endpoints........................ 0.000001
D2 deviation from original curve................... 0.016795
```

- PSK-method

```
D1 deviation from endpoints........................ 0.000001
D2 deviation from original curve................... 0.019622
```

3.1.4.4 Numerical output of fairness measures

- EH-method

```
L21         ............................................ 10.59705205243
L22         ............................................ 0.295364829224
L23         ............................................ 0.052726420393
L2kappa     ............................................ 0.295302100447
L2kappa'    ............................................ 0.028211332014
L2tau       ............................................ 46.37708728451
max_kappa   ............................................ 0.376262363052
#mono_kappa ............................................ 4
max_tau     ............................................ 28.83630034527
#sign_tau   ............................................ 2
```

- PSK-method

```
L21          ............................................   10.59705243121
L22          ............................................    0.296936789770
L23          ............................................    0.088015878548
L2kappa      ............................................    0.296222467139
L2kappa'     ............................................    0.032798838991
L2tau        ............................................   31.51820828323
max_kappa    ............................................    0.383224876170
#mono_kappa  ............................................    6
max_tau      ............................................   19.15206493278
#sign_tau    ............................................    1
```

3.1.4.5 Graphical Output See Figs 3.1.4a-3.1.4c, associated with the EH-method, and Figs. 3.1.4d-3.1.4f, associated with the PSK-method.

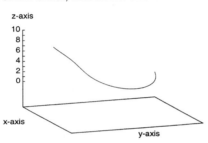

Figure 3.1.4a Chine3 - curve (KCS, EH)

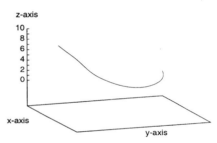

Figure 3.1.4b Chine3 - curve (KCS, PSK)

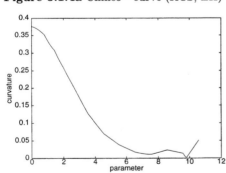

Figure 3.1.4c Chine3 - curvature plot (KCS, EH)

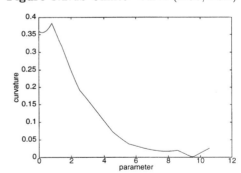

Figure 3.1.4d Chine3 - curvature plot (KCS, PSK)

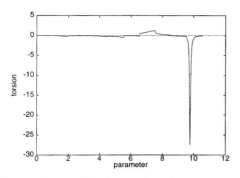

Figure 3.1.4e Chine3 - torsion plot (KCS, EH)

Figure 3.1.4f Chine3 - torsion plot (KCS, PSK)

3.1.5 SUMMARY

Participants

NTUA   K.G. Pigounakis, P.D. Kaklis
TUDH   J. Hadenfeld, J. Hoschek
KCS   I. Applegarth

Numerical output

	L_{21}	L_{22}	L_{23}	$L_{2\kappa}$	$L_{2\kappa'}$	$L_{2\tau}$
INIT	10.5944	0.3202	0.3776	0.3178	0.1010	258.3512
NTUA	10.5851	0.2973	0.0536	0.2939	0.0280	11.7981
TUDH	10.5919	0.2789	0.0286	0.2740	0.0274	0.6543
KCS EH	10.5971	0.2954	0.0527	0.2953	0.0282	46.3771
KCS PSK	10.5971	0.2969	0.0880	0.2962	0.0327	31.5182

	max_κ	$\#mono_\kappa$	max_τ	$\#sign_\tau$
INIT	0.3934	14	134.6233	8
NTUA	0.4112	2	7.4782	4
TUDH	0.3446	3	0.9169	2
KCS EH	0.3763	4	28.8363	2
KCS PSK	0.3832	6	19.1521	1

3.2 Hood - Benchmark

3.2.1 Problem (INIT)

3.2.1.1 Description The surface is a bi-cubic B−spline surface stemming from digitized points from an auto-model hood (units are in millimeters). The data supplier is Daimler Benz AG.

3.2.1.2 Constraints At the boundaries of the surface, G^0 conditions should be maintained. The maximum deviation from the original surface should be less than 1mm.

3.2.1.3 Numerical output of fairness measures

```
L22          ........................................ 340.307838738772
L23          ........................................ 6.39339503171064
L2kappa1,2   ........................................ 1.69447531208431
max|kappa1+2| ....................................... 0.03243892472410
maxK         ........................................ 0.00017526101542
maxH         ........................................ 0.01621946236205
```

3.2.1.4 Graphical Output See Fig. 3.2.1 in the **Appendix: Colour Plates**.

3.2.2 National Technical University of Athens (NTUA)

3.2.2.1 Name and Address T.P. Gerostathis, G.D. Koras, P.D. Kaklis
Department of Naval Architecture and Marine Engineering
National Technical University of Athens
Athens, GREECE
{theo,gkoras,kaklis}@deslab.ntua.gr

3.2.2.2 Method description The method used for surface fairing is the unconstrained minimization of two of the Roulier & Rando metrics, namely the *rolling* and *rounding metric*, as described in [14, 13]. Routine *E04UCF* of *NAg* is employed to solve the non-linear programming problem. The problem is solved in a series of interactive steps. In each one, fairing is focused on a different area on the surface and the control points that have a stronger effect on that area, are chosen to be the free variables.

3.2.2.3 Numerical output of constraints

```
D1 deviation from boundary curves.............. 0.00
D2 deviation from original surface............. 0.708
```

3.2.2.4 Numerical output of fairness measures

```
L22          ................................. 336.50
L23          ................................. 5.6580
L2kappa1,2   ................................. 1.6610
max|kappa1+2| ................................ 0.0318
```

```
maxK  ..................................  0.0002
maxH  ..................................  0.0159
```

3.2.2.5 Graphical Output   See Fig. 3.2.2 in the **Appendix: Colour Plates.**

3.2.3   Technical University of Berlin (TUB)

3.2.3.1 Name and Address   H. Nowacki, G. Westgaard, J. Heimann
Ship-Design Section
Institute of Naval Architecture, Marine and Ocean Engineering
Technical University of Berlin
Berlin, GERMANY
{nowacki,westgaard,heimann}@ism.tu-berlin.de
Date: 22.7.97

3.2.3.2 Method description   The following four (4) alternative measures have been employed for fairing the initially given surface:

```
C1: L22
C2: L23
C3: int(|Suuu|**2 + |Suuv|**2 + |Suvv|**2 + |Svvv|**2)
C5: int(variation of curvature) (parameterization independent)
```

3.2.3.3 Numerical output of constraints

- C1-criterion

```
D1 deviation from boundary curves..............  0.00
D2 deviation from original surface.............  3.894
```

- C2-criterion

```
D1 deviation from boundary curves..............  0.00
D2 deviation from original surface.............  0.555
```

- C3-criterion

```
D1 deviation from boundary curves..............  0.00
D2 deviation from original surface.............  1.119
```

- C5-criterion

```
D1 deviation from boundary curves..............  0.00
D2 deviation from original surface.............  0.824
```

3.2.3.4 Numerical output of fairness measures

- C1-criterion

```
L22           ................................. 185.6477990889491
L23           ................................. 22.64048648621042
L2kappa1,2    ................................. 0.937972980833186
max|kappa1+2| ................................. 0.060490142251391
maxK          ................................. 0.000860137307272
maxH          ................................. 0.019623278852794
```

- C2-criterion

```
L22           ................................. 317.2261931838422
L23           ................................. 1.469813534014230
L2kappa1,2    ................................. 1.595022638990447
max|kappa1+2| ................................. 0.033555827544641
maxK          ................................. 0.000186725347780
maxH          ................................. 0.016777913772320
```

- C3-criterion

```
L22           ................................. 210.9823710203147
L23           ................................. 23.88809987347808
L2kappa1,2    ................................. 1.450786509906652
max|kappa1+2| ................................. 0.029563299725852
maxK          ................................. 0.000115405521213
maxH          ................................. 0.014781649862926
```

- C5-criterion

```
L22           ................................. 268.3897897079916
L23           ................................. 45.47221866134472
L2kappa1,2    ................................. 1.770688146708318
max|kappa1+2| ................................. 0.032142264244374
maxK          ................................. 0.000171121405912
maxH          ................................. 0.016071132122187
```

3.2.3.5 Graphical Output   See Figs. 3.2.3Ci_a-c, i= 1, 2, 3, 5, in the **Appendix: Colour Plates**.

3.2.3.6 Comments

- The boundary constraints have been fulfilled.
- CPU-Time / fairing run (averaged): 30 sec.

3.2.4 Technical University of Darmstadt; Dietz (TUDD)

3.2.4.1 Name and Address Ulrich Dietz, Josef Hoschek
Fachbereich Mathematik, AG3
Technical University of Darmstadt
udietz@mathematik.tu-darmstadt.de
Date: 4.6.97

3.2.4.2 Method description A fitting, rather then a fairing, algorithm is applied to the given data set. In a first step, nine hundred (900) points are sampled from the original surface. Then the control points are recomputed by fitting a B-spline surface with the original orders and knot vectors to the sampled points. To ensure continuity at the surface boundaries two rows of control points are fixed on each boundary. The fitting algorithm combines a least squares fit with the minimization of a simplified bending energy and point parameter correction. In an iteration loop a sequence of B-spline surfaces is calculated by reducing the influence of the energy term until the tolerance constraints are satisfied. The method is described, in detail, in [9, 1].
Limitation: If the original surface is of complex shape the faired surface may have a patch structure very different from the original one.

3.2.4.3 Numerical output of constraints

```
D1 deviation from boundary curves.............. 0.00
D2 deviation from original surface............. 0.321
```

3.2.4.4 Numerical output of fairness measures

```
L22 ................................... 232.778145416815
L23 ................................... 20.3518196281420
L2kappa1,2 ............................ 1.63165885967778
max_kappa1,2 .......................... 0.03243888553067
maxK .................................. 0.00017526071691
maxH .................................. 0.01621944276533
```

3.2.4.5 Graphical Output See Figs 3.2.4 in the **Appendix: Colour Plates**.

3.2.4.6 Comments

- The faired surface is convex *(editorial remark: the Gaussian curvature plot in Fig. 3.2.4b exhibits two tiny regions of convexity failure)*.

- The faired surface is holding C^1 boundary conditions (G^0 required).

- The error tolerance of 1mm is fulfilled.

- One iteration costs 1.2 seconds of CPU-time on a 100MHz HP-workstation

3.2.5 Technical University of Darmstadt; Hadenfeld (TUDH)

3.2.5.1 Name and Address Jan Hadenfeld, Josef Hoschek
Fachbereich Mathematik, AG3
Technical University of Darmstadt
hadenfeld@mathematik.tu-darmstadt.de

3.2.5.2 Method description The employed methods for fairing B-spline curves and surfaces are described, in detail, in [2, 7, 8]. More specifically, for the *Chine3* curve, the integral:

$$E_3 = \int (\mathbf{x}'''(t))^2 \, dt, \tag{1}$$

and for the *Hood* surface, the (simplified) thin-plate energy:

$$\Pi = L_{22} = \iint_A \left(\mathbf{X}_{uu}^2 + 2\,\mathbf{X}_{uv}^2 + \mathbf{X}_{vv}^2 \right) du\, dv \tag{2}$$

are minimized via an iterative method under the constraint that the maximum deviation from the original curve is less than a given tolerance.
The results were calculated on a PC with a Pentium Pro 200MHz processor and in the context of the operating system *Linux*.

3.2.5.3 Numerical output of constraints

```
D1 deviation from boundary curves.............. 0.00
D2 deviation from original surface............. 0.433
```

3.2.5.4 Numerical output of fairness measures

```
L22      ...................................... 274.4202
L23      ...................................... 13.78973
L2kappa1,2 .................................... 1.642207
max|kappa1+2| ................................. 0.032439
maxK     ...................................... 0.000175
maxH     ...................................... 0.016219
```

3.2.5.5 Graphical Output See Fig. 3.2.5 in the **Appendix: Colour Plates**.

3.2.5.6 Comments

- CPU-time: 0.16 sec.

3.2.6 DAIMLER-BENZ AG (DBAG)

3.2.6.1 Name and Address Steffen Wahl, Ekkehart Kaufmann
Abt. EP/QDF HPC: G270
Daimler Benz AG
Sindelfingen, GERMANY
steffen.wahl@icem.de

3.2.6.2 Method description In the selected area, for each u-isoparametric line, a cubic segment with minimal curvature is created. The original u-isoparametric curve is then faired according to this optimal segment with respect to the given tolerance.

3.2.6.3 Numerical output of constraints

```
D1 deviation from boundary curves.............. 0.00
D2 deviation from original surface............. 0.723
```

3.2.6.4 Numerical output of fairness measures

```
L22 ........................................... 345.20
L23 ........................................... 3.6550
L2kappa1,2 .................................... 1.8430
max_kappa1,2 .................................. 0.0315
maxK .......................................... 0.0002
maxH .......................................... 0.0158
```

3.2.6.5 Graphical Output See Fig. 3.2.6 in the **Appendix: Colour Plates**.

3.2.7 Kockums Computer Systems (KCS) UK, Ltd

3.2.7.1 Name and Address Ian Applegarth
Kockums Computer Systems (KCS) UK, Ltd
UNITED, KINGDOM
iana@kcs.co.uk

3.2.7.2 Method description The employed method is based on a paper by Hadenfeld [8], dealing with the fairing of B-Spline curves and surfaces. In the case of surfaces, fairing is achieved through the minimization of an approximation to the thin plate strain energy, i.e, the fairness measure L_{22}. This approach has been extended to handle NURBS surfaces.

3.2.7.3 Numerical output of constraints

```
D1 deviation from boundary curves.............. 0.000000
D2 deviation from original surface............. 0.384001
```

3.2.7.4 Numerical output of fairness measures

```
L22.................................    244.19335058583
L23.................................    17.846739493797
L2kappa1,2..........................    1.6365827147999
max|kappa1+2|  .....................    3.2438872346424E-02
maxK................................    1.7526061640347E-04
maxH................................    1.6219436173212E-02
```

3.2.7.5 Graphical Output See Fig. 3.2.7 in the **Appendix: Colour Plates**.

3.2.8 SUMMARY

Participants

NTUA T.P. Gerostathis, G.D. Koras, P.D. Kaklis
TUB   H. Nowacki, G. Westgaard, J. Heimann
TUDD U. Dietz, J. Hoschek
TUDH J. Hadenfeld, J. Hoschek
DBAG S. Wahl, E. Kaufmann
KCS   I. Applegarth

Numerical output

	D1	D2	L_{22}	L_{23}	$L_{2\kappa_{1,2}}$	$max|\kappa_{1+2}|$	$maxK$	$maxH$
INIT	-	-	340.3	6.393	1.694	0.03244	0.0001753	0.01622
NTUA	0.00	0.708	336.5	5.658	1.661	0.03183	0.0001682	0.01592
TUB C1	0.00	3.894	185.6	22.640	0.938	0.06049	0.0008601	0.01962
TUB C2	0.00	0.555	317.2	1.470	1.595	0.03356	0.0001867	0.01678
TUB C3	0.00	1.119	211.0	23.888	1.451	0.02956	0.0001154	0.01478
TUB C4	0.00	0.555	317.2	1.470	1.595	0.03356	0.0001867	0.01678
TUB C5	0.00	0.824	268.4	45.472	1.771	0.03214	0.0001711	0.01607
TUDD	0.00	0.321	232.8	20.352	1.632	0.03243	0.0001753	0.01622
TUDH	0.00	0.433	240.2	47.940	1.419	0.05301	0.0006555	0.01939
DBAG	0.00	0.723	345.2	3.655	1.843	0.03153	0.0001643	0.01577
KCS	0.00	0.384	244.2	17.84	1.637	0.03244	0.0001752	0.01621

4 Benchmark in the Area of Shape-Constrained Approximation

4.1 Lens - Benchmark

4.1.1 Problem (INIT)

4.1.1.1 Description The point-set consists of 613 points and has been provided by the University of Erlangen-Nürnberg; see Fig. 4.1.1.

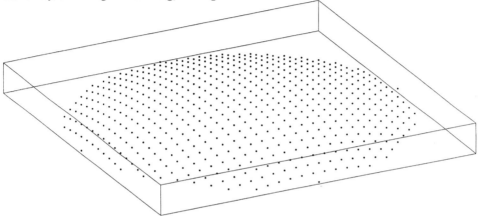

Figure 4.1.1 Lens - data (INIT)

4.1.1.2 Constraints The given point-set should be approximated by a convex B-spline surface with the maximal distance error E as low as possible (in any case $E \leq 0.005$). The data volume should be also kept small. The reflection lines or the isophotes should be well distributed, while the mean curvature (or its derivatives) should be as stable as possible.

4.1.2 Technical University of Darmstadt (TUD)

4.1.2.1 Name and Address Ulrich Dietz, Josef Hoschek
Fachbereich Mathematik, AG3
Technical University of Darmstadt
Darmstadt, GERMANY
udietz@mathematik.tu-darmstadt.de
Date: 3.6.97

4.1.2.2 Method description The employed fitting method calculates, within a prescribed error tolerance, the smoothest tensor-product B-spline surface to a set of scattered points. An initial parameterization of the points is taken from the best fitting plane. Then an iteration loop is started, where linear fitting steps alternate with point parameter optimizations. The fitting steps combine the minimization of the least squares sum and the parametric bending energy. Successively, a sequence of fitting B-spline surfaces is calculated

by reducing the influence of the energy term in each iteration step. This is done until the tolerance constraints are satisfied. The method is described, in detail, in [9, 1].
Limitations: The method cannot guarantee that the generated surface satisfies any convexity conditions.

4.1.2.3 Numerical output of constraints

```
C convex surface............................ yes
E maximal distance error................... 0.04903
N number of control points.................. 16
```

4.1.2.4 Numerical output of fairness measures

```
L22         ........................... 3846.73078680015
L23         ........................... 1045.60197474940
L2kappa1,2  ........................... 1.07713431143388
max|kappa1+2| ......................... 0.03069374586817
maxK        ........................... 0.00023351176414
maxH        ........................... 0.01534687293408
```

4.1.2.5 Graphical Output See Fig. 4.1.2 in the **Appendix: Colour Plates**.

4.1.2.6 Comments The employed hardware platform was a HP workstation. Number of iterations: 10, CPU-time: 4.22 sec.

4.1.3 University of Erlangen-Nürnberg (U_EN)

4.1.3.1 Name and Address G. Greiner, K. Hormann
Lehrstuhl für Graphische Datenverarbeitung
University of Erlangen-Nürnberg
Erlangen-Tennenlohe, GERMANY
{greiner,kihorman}@informatik.uni-erlangen.de
Date: 25.7.97

4.1.3.2 Method description The data set was approximated by a parametric bicubic B-spline, that minimizes the least squares fairness functional:

$$\mathcal{J}(F) = \omega \, J_{Fair}(F) + J_{LS}(F),$$

with

$$J_{Fair} = \underbrace{\int_\Omega (F_{uu}^2 + 2F_{uv}^2 + F_{vv}^2) \, du \, dv}_{\text{Simple Thin Plate Energy}} + \int_\Omega (\alpha F_{uuu}^2 + \beta F_{uuv}^2 + \gamma F_{uvv}^2 + \delta F_{vvv}^2) \, du \, dv,$$

and

$$J_{LS} = \sum_i \|F(T_i) - P_i\|^2.$$

The parameter values $\{T_i\}$, corresponding to the data points $\{P_i\}$, were taken as the projection of the data points into the (x, y)-plane and were iteratively improved by the parameter correction described in [10].

For this data set, bi-cubic B-splines have been used, with 5×5 control points, the optimal parameters were found to be:

$$\omega = 0.000105, \quad \alpha = 2, \quad \beta = 4, \quad \gamma = 4, \quad \delta = 2,$$

while the process of approximation/parameter correction was iterated 15 times. The method is described, in detail, in [3, 4, 5, 6].

4.1.3.3 Numerical output of constraints

```
C convex surface.............................. yes
E maximal distance error...................... 0.049621
N number of control points.................... 16
```

4.1.3.4 Numerical output of fairness measures

```
L22       ......................................... 3055.284864581587
L23       ......................................... 268.3855939308837
L2kappa1,2 ....................................... 0.943832557016675
max_kappa1,2 ..................................... 0.028791207479244
maxK      ......................................... 0.000207067345977
maxH      ......................................... 0.014395603739622
```

4.1.3.5 Graphical Output   See Fig. 4.1.3 in the **Appendix: Colour Plates**.

4.1.3.6 Comments   The approximation process took approximately 0.25 seconds CPU-time on a Silicon Graphics ONYX.

4.1.4   University of Kaiserslautern (U_K)

4.1.4.1 Name and Address   M. Latz, S. Heinz
FB Informatik
University of Kaiserslautern
Kaiserslautern, GERMANY
heinz@informatik.uni-kl.de
Date: 4.7.97

4.1.4.2 Method description   Variational technique implemented in the context of the commercial CAD System CATIA SOLUTIONS V4 RELEASE 1.6 FR 4.1.

4.1.4.3 Numerical output of constraints

```
C convex surface.............................. yes
E maximal distance error...................... 0.038299
N number of control points.................... 36
```

4.1.4.4 Numerical output of fairness measures

```
L22 ............................................ 4597
L23 ............................................ 1371
L2kappa1,2 .................................... 1.11900
max|kappa1+2| ................................. 0.02960
maxK .......................................... 0.00022
maxH .......................................... 0.01480
```

4.1.4.5 Graphical Output See Fig. 4.1.4 in the **Appendix: Colour Plates**.

4.1.4.6 Comments

- Performance: 30 seconds on a IBM RS6000

4.1.5 Summary

Participants

TUD $\quad$ $U.\ Dietz,\ J.\ Hoschek$
U_EN $\quad$ $G.\ Greiner,\ K.\ Hormann$
U_K $\quad$ $M.\ Latz,\ S.\ Heinz$

Numerical output

	C	E	N	L_{22}	L_{23}	$L_{2\kappa_{1,2}}$	$max\|\kappa_{1+2}\|$	$maxK$	$maxH$
TUD	yes	0.04903	16	3847	1046	1.077	0.03069	0.0002335	0.01535
U_EN	yes	0.04962	16	3055	268	0.944	0.02879	0.0002071	0.01440
U_K	yes	0.03830	36	4597	1371	1.119	0.02960	0.0002189	0.01480

4.2 WaterLines - Benchmark

4.2.1 Problem (INIT)

4.2.1.1 Description The data set is a point-set, measured form the end-surface region of the fore part of a ship, that is decomposed into nine (9) subsets. Each subset, referred to as water line, lies on the intersection of the hull surface with a plane of constant $z-$level and consists of sixty (60) points. The created surface should be a bi-cubic B-spline surface with Gaussian curvature of constant sign. The data set has been supplied by KCS UK, Ltd.

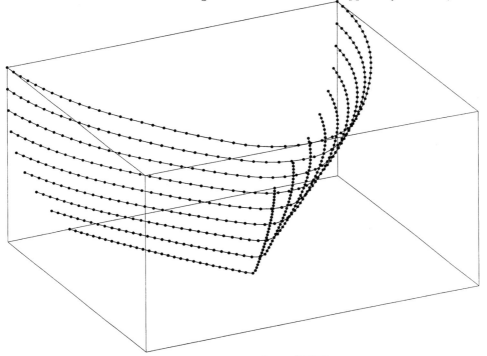

Figure 4.2.1 Water Lines (INIT)

4.2.1.2 Constraints

- positive Gaussian curvature
- tangent constraints (symmetry)

4.2.2 University of Erlangen-Nürnberg (U_EN)

4.2.2.1 Name and Address G. Greiner, K. Hormann
Lehrstuhl für Graphische Datenverarbeitung
University of Erlangen-Nürnberg
Erlangen-Tennenlohe, GERMANY
{greiner,kihorman}@informatik.uni-erlangen.de

Date: 25.7.97

4.2.2.2 Method description The data set was approximated by a parametric bi-cubic B-spline that minimizes the least squares fairness functional:

$$\mathcal{J}(F) = \omega\, J_{Fair}(F) + J_{LS}(F),$$

with

$$J_{Fair} = \underbrace{\int_\Omega (F_{uu}^2 + 2F_{uv}^2 + F_{vv}^2)\, du\, dv}_{\text{Simple Thin Plate Energy}} + \int_\Omega (\alpha F_{uuu}^2 + \beta F_{uuv}^2 + \gamma F_{uvv}^2 + \delta F_{vvv}^2)\, du\, dv,$$

and

$$J_{LS} = \sum_i \|F(T_i) - P_i\|^2.$$

The parameter values $\{T_i\}$, corresponding to the data points $\{P_i\}$, were taken as a regular 59×9 grid in the parameter space and were iteratively improved by the parameter correction technique described in [10].

For this data set, bi-cubic B-Splines have been used, with 7×7 control points, the optimal parameters were found to be:

$$\omega = 0.0001, \qquad \alpha = 20, \qquad \beta = 5, \qquad \gamma = 5, \qquad \delta = 500,$$

while the process of approximation/parameter correction was iterated 15 times.

4.2.2.3 Numerical output of constraints

```
P positive Gaussian curvature................ yes
E maximal distance error..................... 0.528
N number of control points................... 49
S tangent constraints (symmetry)............. yes
```

4.2.2.4 Numerical output of fairness measures

```
L22       ............................................ 8314.873747768757
L23       ............................................ 3818.712500429995
L2kappa1,2 ........................................... 5.655486186958465
max|kappa1+2| ........................................ 0.993707678055808
maxK      ............................................ 0.002584355053979
maxH      ............................................ 0.496853839027904
```

4.2.2.5 Graphical Output See Fig. 4.2.2 in the **Appendix: Colour Plates**.

4.2.2.6 Comments

- The approximation process took approximately 0.3 seconds CPU-time on a Silicon Graphics ONYX.

4.2.3 University of Kaiserslautern (U_K)

4.2.3.1 Name and Address S. Heinz, M. Latz
FB Informatik
University of Kaiserslautern
Kaiserslautern, GERMANY
heinz@informatik.uni-kl.de
Date: 27.6.97

4.2.3.2 Method description Variational technique implemented in the context of the commercial CAD system CATIA SOLUTIONS V4 RELEASE 1.6 FR 4.1.

4.2.3.3 Numerical output of constraints

- Smoothing weight 20 (SW20)

```
P positive Gaussian curvature................ no
E maximal distance error..................... 0.269761
N number of control points................... 36
S tangent constraints (symmetry)............. no
```

- Smoothing weight 50 (SW50)

```
P positive Gaussian curvature................ no
E maximal distance error..................... 0.26188
N number of control points................... 36
S tangent constraints (symmetry)............. no
```

4.2.3.4 Numerical output of fairness measures

- Smoothing weight 20 (SW20)

```
L22       ............................................. 466.1
L23       ............................................. 1231
L2kappa1,2 ........................................ 1.336
max|kappa1+2| ..................................... 0.2362
maxK      ............................................. 0.01187
maxH      ............................................. 0.1077
```

- Smoothing weight 50 (SW50)

```
L22       ............................................. 296.8
L23       ............................................. 1363
L2kappa1,2 ........................................ 1.020
max|kappa1+2| ..................................... 0.2125
maxK      ............................................. 0.008561
maxH      ............................................. 0.09032
```

4.2.3.5 Graphical Output For Smoothing Weight equal to 20 (SW20), see Figs. 4.2.3a-c, while for SW=50, see Figs. 4.2.3.d-f in the **Appendix: Colour Plates**.

4.2.3.6 Comments

- Performance: 15 seconds on an IBM RS6000 workstation.

4.2.4 Summary

Participants

U_EN G. Greiner, K. Hormann
U_K   M. Latz, S. Heinz

Numerical output

	P	E	N	S	L_{22}	L_{23}	$L_{2\kappa_{1,2}}$	$max|\kappa_{1+2}|$	$maxK$	$maxH$
U_EN	yes	0.528	49	yes	8315	3819	5.655	0.9937	0.002584	0.4969
U_K SW20	no	0.2698	36	no	466.1	1231	1.336	0.2362	0.011870	0.1077
U_K SW50	no	0.2619	36	no	296.8	1363	1.020	0.2125	0.008561	0.0903

5 Comparative Remarks

Based on the tabulated results, collected in the **Summary** subsection of each benchmark problem, as well as the **Graphical Output** accompanying each benchmark contribution, we hereafter provide comparative comments on the quality of fairness achieved by the received contributions. Before proceeding, however, two clarifications are due. First, the ensuing remarks may be taken as objective to the extent that the interested reader accepts the set of fairness measures, adopted by the Steering Committee of FAIRSHAPE, as a satisfactory tool for evaluating quantitatively the fairness of a curve/surface. Second, the remarks are free from any attempt to explain the observed differences in performance, for that it would require, in most cases, a rather intrinsic comparison for each pair of the employed techniques.

5.1 Chair-Benchmark (§2)

The performance of the technique, proposed by the University of Florence (U_FIR), is definitely better than that achieved by the technique proposed by the National Technical University of Athens (NTUA). More specifically, the U_FIR technique yields a three-dimensional interpolant that is considerably fairer than the NTUA technique, in terms of the integral measures $L_{2\kappa}$, $L_{2\kappa'}$, $L_{2\tau}$, and the absolute curvature and torsion maxima, max_κ and max_τ, as well. The opposite picture is observed for the integral measures L_{21}, L_{22} and L_{23}, which is not, however, taken into account for these functionals are parameterization depending. Finally, the curve, provided by the NTUA technique, is slightly better in terms of the number of monotonic segments of curvature ($\#mono\_\kappa$) and the number of sign changes in the torsion plot ($\#sign\_\tau$).

5.2 Chine3-Benchmark (§3.1)

The performance of the technique, proposed by the Technical University of Darmstadt (TUDH), is definitely the best among the four (4) contributions to this benchmark problem. The TUDH curve is the fairest in terms of the integral measures $L_{2\kappa}$, $L_{2\kappa'}$, $L_{2\tau}$, and the absolute curvature and torsion maxima. Regarding the number of monotonic segments of curvature and the number of sign changes in torsion plot, the best (smallest) figures are achieved by the NTUA and KCS-PSK technique, respectively. The latter is one of the two (2) contributions received from Kockums Computer Systems UK, Ltd (KCS). The fact that the NTUA technique yields the best, from the monotonicity point of view, curvature plot, follows also by simply comparing the corresponding curvature plots; compare Fig. 3.1.2b versus Figs. 3.1.3c and 3.1.4c,d. Finally, a difference that deserves further investigation: while torsion is mainly negative for all four (4) compared techniques, its maxima is negative for the NTUA and KCS-EH techniques (see Figs. 3.1.2c, 3.1.4e) and positive for the TUDH and KCS-PSK techniques (see Figs. 3.1.3d, 3.1.4f).

5.3 Hood-Benchmark (§3.2)

This has been the most popular among the problems of the Benchmark Experiment, having received nine (9) contributions from five (5) FAIRSHAPE institutions. At first, we have to exclude from the comparison poll those techniques, that fail to fulfill the constraint: *"the maximum deviation from the original surface should be less than 1mm"*; see §3.2.1.2. Thus, the ensuing remarks are not influenced from the output of the TUB-C1 and TUB-C3 techniques, which are two (2) among the four (4) techniques proposed by the Technical University of Berlin (TUB).

Nevertheless, it is still hard to attribute the best-performance characterization to a single among the remaining seven (7) techniques. We thus confine ourselves to simply marking the technique proposed by Daimler Benz AG (DBAG), whose output is the fairest one with regard to the fairness measures $max|\kappa_{1+2}|$, $maxK$ and $maxH$, while it achieves to be second in the ranking list for the integral measure L_{23}.

Furthermore, if we consider the criteria of convexity and smooth distribution of isophotes, as worth-noticing criteria by themselves, then we can say:

- The NTUA technique is definitely the best, providing a locally convex surface (positive Gaussian curvature) over the whole parametric domain; compare Fig. 3.2.2b versus Figs. 3.2.3C2_b, 3.2.3.C5_b, 3.2.4b, 3.2.5b, 3.2.6b and 3.2.7b in the **Appendix: Colour Plates**.

- The surfaces, provided by the two (2) techniques proposed by the Technical University of Darmstadt, denoted as the TUDD and TUDH techniques, seem to exhibit the smoothest isophote distributions; compare Figs. 3.2.4a and 3.2.5a versus Figs. 3.2.2a, 3.2.3C2_a, 3.2.3C4_a, 3.2.3C5_a, 3.2.6a and 3.2.7a in the **Appendix: Colour Plates**.

5.4   Lens-Benchmark (§4.1)

The performance of the technique, proposed by the University of Erlangen-Nürnberg (U_EN), is definitely the best among the three (3) received contributions to this benchmark problem. The U_EN surface not only satisfies the convexity and deviation constraints, imposed by the data provider (see §4.1.1.2), but it is also the fairest one with regard to the whole set of the adopted fairness measures. In addition, it conforms with the recommendation to keep the data volume as small as possible, using only sixteen (16) control points for the approximating surface. As for the constraint that *"the mean curvature (or its derivatives) should be as stable as possible"* (see §4.1.1.2), it seems that the mean-curvature distribution, achieved by the technique of the University of Kaiserslautern (U_K), is the most uniformly distributed, though quite close to the zero level, in comparison with that obtained by U_EN and the Technical University of Darmstadt (TUD); compare Fig. 4.1.4c with Figs. 4.1.2c and 4.1.3c in the **Appendix: Colour Plates**.

5.5   WaterLines-Benchmark (§4.2)

Among the three (3) received contributions, only that, provided by the University of Erlangen-Nürnberg (U_EN), fulfills the constraints described in §4.2.1.2. Thus, there exists no common basis for rigorously comparing the output of the U_EN technique with the other two (2) contributions originating from the University of Kaiserslautern (U_K), denoted as the U_K-SW20 and U_K-SW50 contribution.

6 Doing It Yourself

The present section enables the interested reader-researcher to test her/his *fairing* and/or *shape-preservation* software versus the data proposed by the FAIRSHAPE Benchmarking Experiment. For this purpose, we have included below in `verbatim` format the FINAL ANNOUNCEMENT of the Benchmarking Experiment, containing the GENERAL RULES OF THE BENCHMARK [1] and all necessary information for downloading the benchmark data (see `Appendix II, III` within the present section) and the accompanying software (see `Appendix I`), as well us uploading her/his contribution to the proper `ftp-site` (see `Appendix IV`).

```
********************************************************************

        FINAL ANNOUNCEMENT: FAIRSHAPE BENCHMARK EXPERIMENT
       "Benchmark on Fair Shape Generation Subject to Constraints"

********************************************************************

BENCHMARK EXPERIMENT COORDINATORS (in parallel)
**************************************************
- Assoc.Prof. Dr.-Ing. P.D. Kaklis, NTUA, Athens
  Email: kaklis@deslab.ntua.gr

- Ian Applegarth, KCS, Wallsend U.K.
  Email: iana@kcs.co.uk

- Steffen Wahl, Daimler Benz AG, Sindelfingen
% Email: steffen_wahl@ep.mbag.sifi.daimlerbenz.com
  Email: steffen.$\!$wahl@icem.de

Any other questions may also be addressed to
  Prof. Horst Nowacki, TU Berlin
  Email: nowacki@cadlab.tu-berlin.de




GENERAL OBJECTIVES
******************
The European Research Network on "Automatic Fairing and Shape-
Preserving Methodologies for CAD/CAM" (European Commission,
DG XII, HCM Framework Programme) is conducting a benchmark experiment
on the generation of fair curves and surfaces. The benchmark
tests will seek to establish a state of the art evaluation of
the quality of fairness achievable by current curve and surface
fairing methodologies in view of different fairness measures
and subject to a variety of geometric constraints. The results
```

---

[1]Please ignore RULE 10!

of the benchmark will be documented and will also be reported
in overview at the FAIRSHAPE Workshop on "Creating Fair and
Shape Preserving Curves and Surfaces" to be held in Berlin on
14-17 September 1997.

SCOPE
*****
The benchmark will comprise several test data sets for curves
and surfaces which are to be treated by the participants' fair
shape generation algorithms. The results shall be analyzed
based on several fairness measures using software to be provided by the FAIRSHAPE Project.

FAIRNESS MEASURES
****
The benchmark will be based on four different fairness measures, each
for curves and surfaces.

For curves:

- Integrals of squared parametric derivatives of first, second
  or third order
- Curvature squared
- Change of curvature squared
- Torsion squared

For surfaces:

- Parametric approximation of strain energy
- Third mixed derivatives (parametric)
- Total curvature expression
- Smallest maximum of absolute curvature

Other fairness measures are optional.

Participants may construct their curves and surfaces by any
desired algorithms, minimizing any or none of the above
measures. However the resulting shapes must be evaluated for
each of the given measures for comparison purposes.

SOFTWARE FOR ANALYZING FAIRNESS MEASURES
****

The latest version of the software for evaluating curve and
surface fairness measures (as outlined above) has been e-mailed
by Mercedes-Benz AG to NTUA.
This software has to be applied uniformly by all benchmark
participants for the evaluation of the fairness measures in
order to ensure a proper comparison of the results archived.
The software has been installed for downloading at the
ftp-site of the Ship-Design Laboratory of NTUA (see Appendix I).

DESCRIPTION OF THE BENCHMARK PROBLEMS/DATA SETS
****

The test data sets are related to three categories of tasks:
- Shape-constrained approximation,
- shape-constrained interpolation,
- fairing subject to discrete constraints, e.g.,
  interpolation constraints.

The ftp-site of the Ship-Design Laboratory of NTUA
now contains the full and latest description of the benchmark
problems accompanied by all the data sets (see Appendix II).

GENERAL RULES OF THE BENCHMARK
****

1. Participants may submit results for any or all of the test
   datasets.

2. In all problems a fair solution is desired (as fair as
   possible under given constraints).

3. Participants use their own algorithms to construct the
   shapes.

4. Fairness measures for the generated shapes are evaluated by
   the participants using FAIRSHAPE standard software
   available from the NTUA ftp-site (see Appendix I).

5. Constraints are also checked by the participants
   (some checking software is available via ftp).

6. Solutions for surfaces should be represented as tensor product surfaces.

7. Spreadsheet Template for submission of solution to one (1) problem:
   7.1 Identification header (in ASCII format)
       - name
       - affiliation
       - email-address
       - date
   7.2 Problem identification (in ASCII format)
       - short title and number
   7.3 Method description (in LaTeX format)
   7.4 Result description (in LaTeX format)
       - numerical output of fairness measures
       - numerical output of constraints, deviations, tolerances
       - comments (optional)
   7.5 Result data in ASCII format, using extended Hadenfeld format (see Appendix III) for the exchange of B-spline curve and surface data and set of points

8. Rules for documenting the results:
   8.1 please describe the method applied
       - limitations
       - reference to literature
       - space of functions used
       - CAD-system used
   8.2 please give an indication of performance of the algorithm (like CPU-time etc)
   8.3 please do not send figures (will be asked for by the coordinators, if necessary)
   8.4 please use MIME/uuencode/etc to compress data
   8.5 please send solutions to the NTUA ftp-site as given in Appendix IV

9. The results will be documented in a publishable report. A summary of all results will be reported at the FAIRSHAPE workshop in September 1997.

10. SUBMISSION Deadline for benchmark results is

```
************************
***   JULY 15th, 1997  ***
************************
```

This date applies uniformly to all participants.

**████████

Appendix I:

The software for analyzing fairness measures has been installed
for downloading by NTUA in the directories:
   "pub/deslab/CAGDFAIRSHAPE_curve_crit"
and
   "pub/deslab/CAGDFAIRSHAPE_surface_crit"
at the ftp-site of the Ship-Design Laboratory of NTUA
which can be reached by anonymous ftp at
   "ftp.deslab.naval.ntua.gr".

Appendix II:

The directory:
   "pub/deslab/CAGD/BERLIN_POTSDAM_workshop"
of the ftp-site of the Ship-Design Laboratory of NTUA
which can be reached by anonymous ftp at
   "ftp.deslab.naval.ntua.gr"
now contains the full and latest description of the benchmark problems
accompanied by the data sets.
To get this description one can do either of the following:

1. Download all "*.tex", "*.ps" and "*.sty" files and
   compile by "tex" and then by "dvips" the file
   "benchmark.tex" or "benchmark_ps_free.tex", which will
   yield the full description of the benchmark problems
   in postscript format with or without the accompanying
   figures.

2. Download and then "uncompress" either of the
   "*.Z" files, which will yield the full description
   of the benchmark problems in postscript format with or
   without the accompanying figures.

Appendix III:

A description of the Hadenfeld format is available
with the following URLs:

http://www.mathematik.th-darmstadt.de/~hadenfeld/beschr-engl.tex
http://www.mathematik.th-darmstadt.de/~hadenfeld/beschr-engl.dvi
http://www.mathematik.th-darmstadt.de/~hadenfeld/beschr-engl.ps).

Visualization of isophotes etc. for this format will be
provided by the benchmark organizers.

Appendix IV:

For security reasons the standard procedure for uploading files in an
anonymous ftp server is to upload them in the directory
   "/pub/incoming"
and immediately inform the system administrator so that he/she can move
the files to the directory they should be in.

So, after you upload the benchmark-results/solutions in
   "/pub/incoming"
at the ftp-site of the Ship-Design Laboratory of NTUA
which can be reached by anonymous ftp at
   "ftp.deslab.naval.ntua.gr",
send an e-mail message to
   "<ftpadmin@deslab.ntua.gr>".

Note that even after you upload the files, "/pub/incoming" will be
apparently empty. The reason for that is that the anonymous user has
write but no read permission in this directory. This is common to
anonymous ftp-servers to avoid data transfer between anonymous users
without the administrator knowing.

7 APPENDIX: Colour Plates

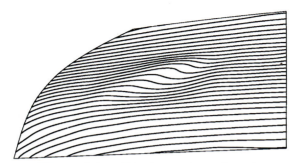

Figure 3.2.1a Hood - Isophotes (INIT)

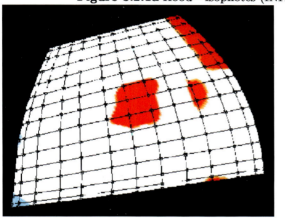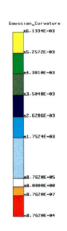

Figure 3.2.1b Hood - Gaussian curvature (INIT)

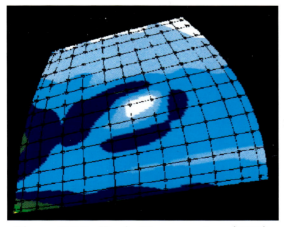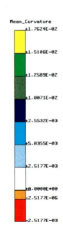

Figure 3.2.1c Hood - Mean curvature (INIT)

7 APPENDIX: Colour Plates 50

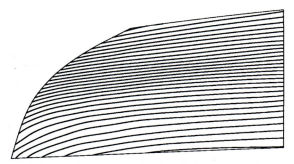

Figure 3.2.2a Hood - Isophotes (NTUA)

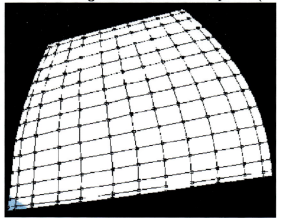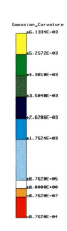

Figure 3.2.2b Hood - Gaussian curvature (NTUA)

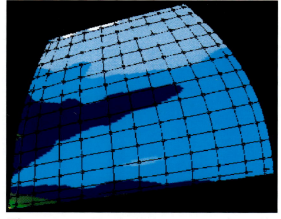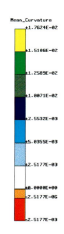

Figure 3.2.2c Hood - Mean curvature (NTUA)

7 APPENDIX: Colour Plates   51

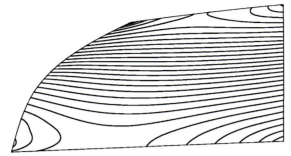

Figure 3.2.3C1_a Hood - Isophotes (TUB, C1-criterion)

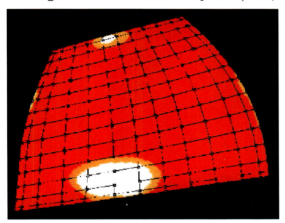

Figure 3.2.3C1_b Hood - Gaussian curvature (TUB, C1-criterion)

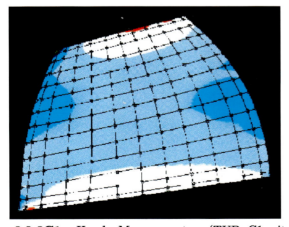
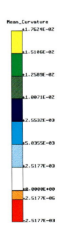

Figure 3.2.3C1_c Hood - Mean curvature (TUB, C1-criterion)

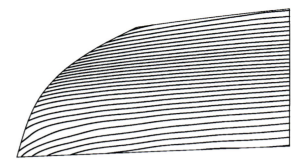

Figure 3.2.3C2_a Hood - Isophotes (TUB, C2-criterion)

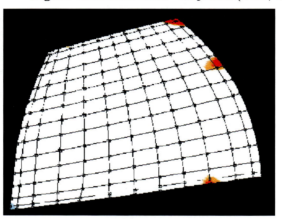

Figure 3.2.3C2_b Hood - Gaussian curvature (TUB, C2-criterion)

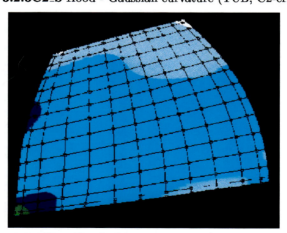

Figure 3.2.3C2_c Hood - Mean curvature (TUB, C2-criterion)

7 APPENDIX: Colour Plates 53

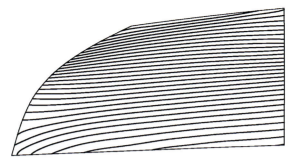

Figure 3.2.3C3_a Hood - Isophotes (TUB C3)

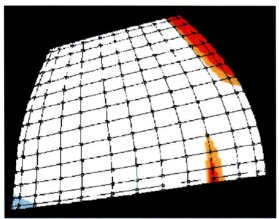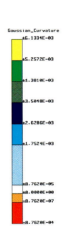

Figure 3.2.3C3_b Hood - Gaussian curvature (TUB, C3-criterion)

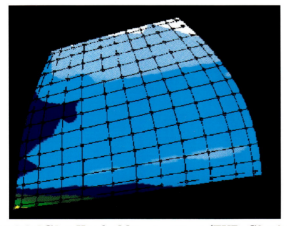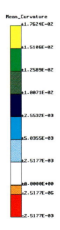

Figure 3.2.3C3_a Hood - Mean curvature (TUB, C3-criterion)

7 APPENDIX: Colour Plates 54

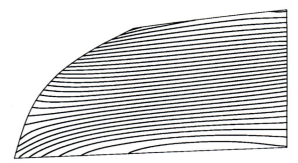

Figure 3.2.3C5_a Hood - Isophotes (TUB, C5-criterion)

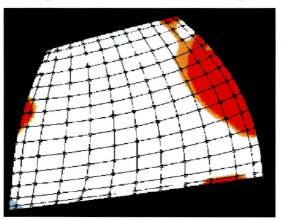

Figure 3.2.3C5_b Hood - Gaussian curvature (TUB, C5-criterion)

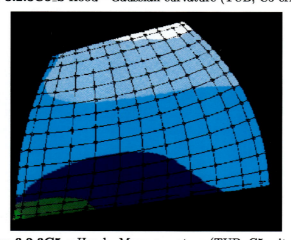

Figure 3.2.3C5_c Hood - Mean curvature (TUB, C5-criterion)

7 APPENDIX: Colour Plates 55

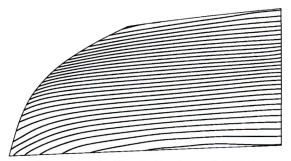

Figure 3.2.4a Hood - Isophotes (TUDD)

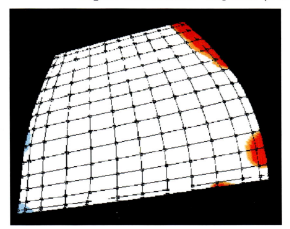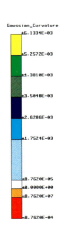

Figure 3.2.4b Hood - Gaussian curvature (TUDD)

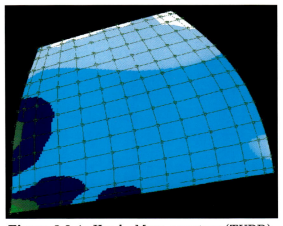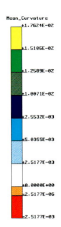

Figure 3.2.4c Hood - Mean curvature (TUDD)

7 APPENDIX: Colour Plates 56

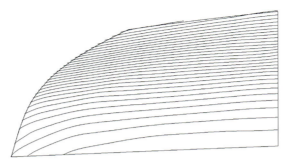

Figure 3.2.5a Hood - Isophotes (TUDH)

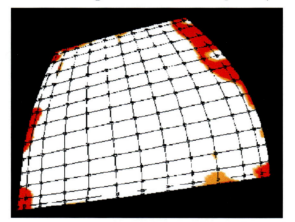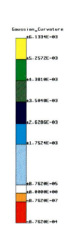

Figure 3.2.5b Hood - Gaussian curvature (TUDH)

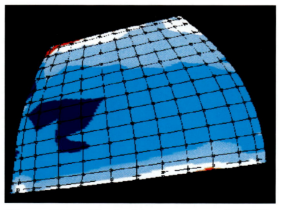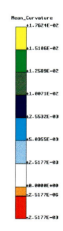

Figure 3.2.5c Hood - Mean curvature (TUDH)

7 APPENDIX: Colour Plates 57

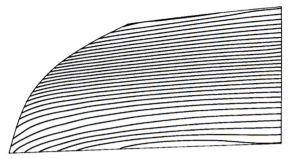

Figure 3.2.6a Hood - Isophotes (DBAG)

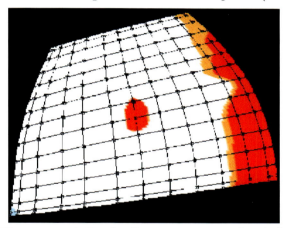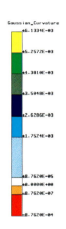

Figure 3.2.6b Hood - Gaussian curvature (DBAG)

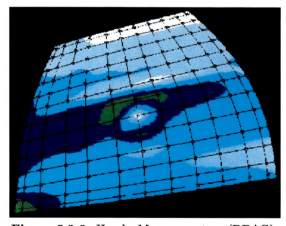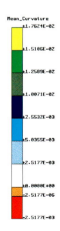

Figure 3.2.6c Hood - Mean curvature (DBAG)

7 APPENDIX: Colour Plates   58

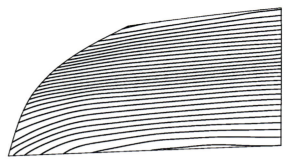

Figure 3.2.7a Hood - Isophotes (KCS)

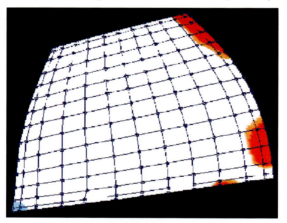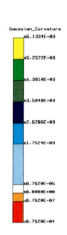

Figure 3.2.7b Hood - Gaussian curvature (KCS)

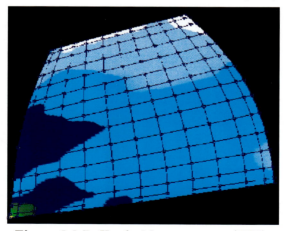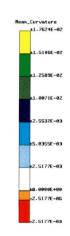

Figure 3.2.7c Hood - Mean curvature (KCS)

7 APPENDIX: Colour Plates   59

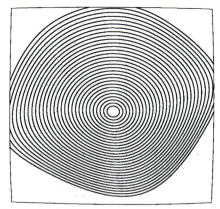

Figure 4.1.2a Lens - Isophotes (TUDD)

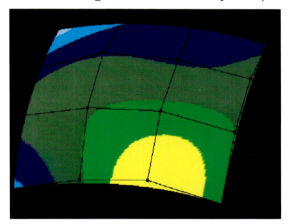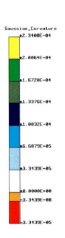

Figure 4.1.2b Lens - Gaussian curvature (TUDD)

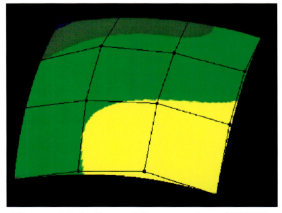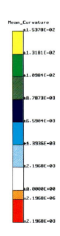

Figure 4.1.2c Lens - Mean curvature (TUDD)

7 APPENDIX: Colour Plates   60

Figure 4.1.3a Lens - Isophotes (U_EN)

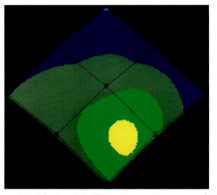
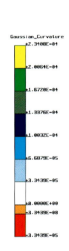

Figure 4.1.3b Lens - Gaussian curvature (U_EN)

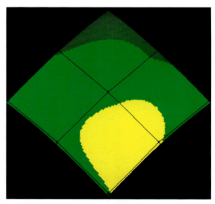
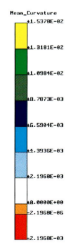

Figure 4.1.3c Lens - Mean curvature (U_EN)

7 APPENDIX: Colour Plates 61

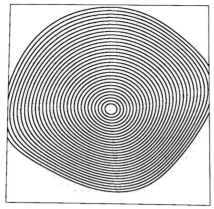

Figure 4.1.4a Lens - Isophotes (U.K)

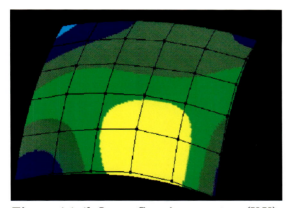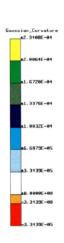

Figure 4.1.4b Lens - Gaussian curvature (U.K)

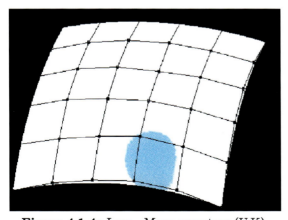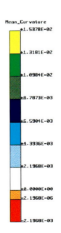

Figure 4.1.4c Lens - Mean curvature (U.K)

7 APPENDIX: Colour Plates   62

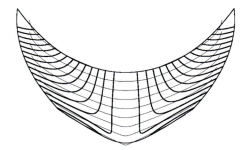

Figure 4.2.2a Water Lines - Isophotes (U_EN)

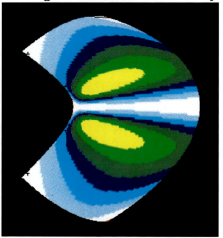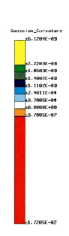

Figure 4.2.2b Water Lines - Gaussian curvature (U_EN)

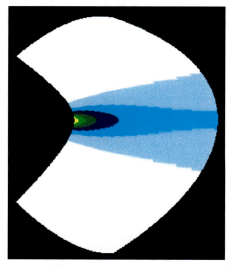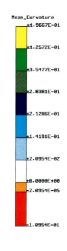

Figure 4.2.2c Water Lines - Mean curvature (U_EN)

7 APPENDIX: Colour Plates   63

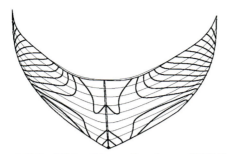

Figure 4.2.3a Water Lines - Isophotes (U_K SW20)

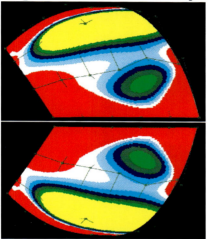 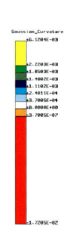

Figure 4.2.2b Water Lines - Gaussian curvature (U_K SW20)

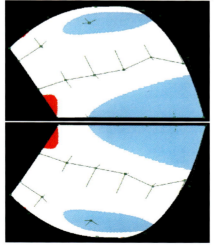 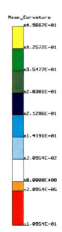

Figure 4.2.2c Water Lines - Mean curvature (U_K SW20)

7 APPENDIX: Colour Plates 64

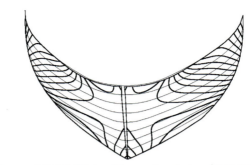

Figure 4.2.2d Water Lines - Isophotes (U.K SW50)

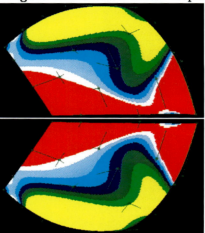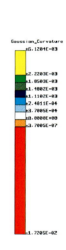

Figure 4.2.2e Water Lines - Gaussian curvature (U.K SW50)

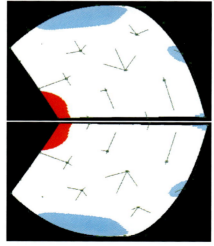

Figure 4.2.2f Water Lines - Mean curvature (U.K SW50)

References

[1] U. Dietz, "B-Spline Approximation with Energy Constraints", In *Advanced Course on FAIRSHAPE*, J. Hoschek and P. D. Kaklis (eds), B.G. Teubner Verlag, Stuttgart, pp. 229-400, 1996.

[2] M. Eck and J. Hadenfeld, "Local Energy Fairing of B-Spline Curves", In *Computing Supplementum*, Vol. 10, G. Farin, H. Hagen and H. Noltemeier (eds), Springer Verlag, pp. 129-147, 1995.

[3] G. Greiner, "Surface construction based on variational principles", In *Wavelets, Images and Surface Fitting*, P. J. Laurent, A. Le Méhauté, L. L. Schumaker (eds), AK Peters, Wellesley, MA, pp. 277–286, 1994.

[4] G. Greiner. "Variational Design and Fairing of Spline Surfaces", *Computer Graphics Forum*, vol. 13, pp. 143–154, 1994.

[5] G. Greiner, J. Loos, W. Wesselink, "Data Dependent Thin Plate Energy and its use in Interactive Surface Modeling", *Computer Graphics Forum*, vol. 15, pp. 175–186, 1996.

[6] G. Greiner, K. Hormann, "Interpolating and Approximating Scattered 3D-data with Hierarchical Tensor Product B-Splines", In *Surface Fitting and Multiresolution Methods*, A. Le Méhauté, C. Rabut, L. L. Schumaker (eds), Vanderbilt University Press, Nashville, TN, pp. 163–172, 1997.

[7] J. Hadenfeld, "Local Energy Fairing of B-Spline Surfaces", In *Mathematical Methods for Curves and Surfaces*, M. Dæhlen, T. Lyche and L. L. Schumaker (eds), Vanderbilt University Press, pp. 203–212, 1995.

[8] J. Hadenfeld, "Fairing of B-Spline Curves and Surfaces", In *Advanced Course on FAIRSHAPE*, J. Hoschek and P. D. Kaklis (eds), B.G. Teubner Verlag, Stuttgart, pp. 59-75, 1996.

[9] J. Hoschek, U. Dietz, "Smooth B-Spline Surface Approximation to Scattered Data", In *Reverse Engineering*, J. Hoschek and W. Dankwort (eds), B.G. Teubner Verlag Stuttgart, pp. 143-152, 1996.

[10] J. Hoschek, D. Lasser, *Grundlagen der Geometrischen Datenverarbeitung*, B. G. Teubner Verlag, Stuttgart, 1989.

[11] P.D. Kaklis and M.I. Karavelas. "Shape Preserving Interpolation in $I\!R^3$", IMA J. Numerical Analysis vol. 17, pp. 373-419, 1997.

[12] K.G. Pigounakis, N.S. Sapidis and P.D. Kaklis, "Fairing Spatial B-Spline Curves", *Journal of Ship Research*, Vol. 40, No. 4, pp. 351-367, 1996.

[13] T. Rando, J. Roulier, "Designing Faired Parametric Surfaces", *Computer-Aided Design*, vol. 23, pp. 492-497, 1991.

[14] J. Roulier, T. Rando, "Measures of Fairness for Curves and Surfaces", In *Designing Fair Curves and Surfaces*, N.S. Sapidis (ed.), SIAM, pp. 75-122, 1994.

Nowacki/Kaklis (Eds.)
Creating Fair and Shape-Preserving Curves and Surfaces

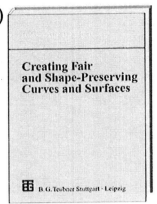

Edited by
Prof. Dr.-Ing.
Horst Nowacki
Technische Universität Berlin
and Prof. Dr.-Ing.
Panagiotis D. Kaklis
National Technical University
of Athens/Greece

1998. 288 pages.
16,2 x 23,5 cm.
Paper DM 59,80
ÖS 437,– / SFr 54,–
ISBN 3-519-02636-8

The creation of free-form curves and surfaces by fairing processes, especially with shape-preservation constraints, is an important task in the geometric and functional design of many products, e. g., car body surfaces, aircraft wings and fuselages, and ship hulls. Rational, quantitative fairness measures are used to compare different shapes and measure the success of a fairing process. Robust and efficient algorithms are needed to secure the geometric quality of technical shapes. The fairing process must also conserve the essential shape character captured in the given data sets.

The results presented in this book are based on an International Workshop held the EU Projekt FAIRSHAPE where a comprehensive overview of state of the art methodologies and recent new developments in fairing and shape-preserving processes for geometric design was compiled. In addition the experiences from a major new benchmark test on diverse algorithms with many practical examples and different fairness criteria are also reported. The book combines direct project results from FAIRSHAPE with other contributions from the scientific community.

Preisänderungen vorbehalten.

 B. G. Teubner Stuttgart · Leipzig
Postfach 80 10 69 · 70510 Stuttgart